Analogies

for the 21st Century

Routledge
Taylor & Francis Group
NEW YORK AND LONDON

Written by **Bonnie Risby**
Illustrated by **Stephanie O'Shaughnessy**

First published in 2008 by Prufrock Press Inc.

Published 2021 by Routledge
605 Third Avenue, New York, NY 10017
2 Park Square, Milton Park, Abingdon, Oxon OX14 4RN

Routledge is an imprint of the Taylor & Francis Group, an informa business

ISBN 13: 978-1-5936-3047-8 (pbk)

DOI: 10.4324/9781003233022

Contents

∎∎∎∎∎∎∎∎∎∎∎∎∎∎∎∎∎∎∎∎∎∎∎∎∎∎∎∎∎∎∎∎∎∎∎∎∎∎∎

Analogies for the 21st Century

All lessons in this book align to the following standards.

Grade Level	Common Core State Standards in ELA-Literacy
Grade 4	L.4.4 Determine or clarify the meaning of unknown and multiple-meaning words and phrases based on grade 4 reading and content, choosing flexibly from a range of strategies.
Grade 5	L.5.5 Demonstrate understanding of figurative language, word relationships, and nuances in word meanings. L.5.6 Acquire and use accurately grade-appropriate general academic and domain-specific words and phrases, including those that signal contrast, addition, and other logical relationships (e.g., however, although, nevertheless, similarly, moreover, in addition).
Grade 6	L.6.5 Demonstrate understanding of figurative language, word relationships, and nuances in word meanings. L.6.6 Acquire and use accurately grade-appropriate general academic and domain-specific words and phrases; gather vocabulary knowledge when considering a word or phrase important to comprehension or expression.
Grade 7	L.7.5 Demonstrate understanding of figurative language, word relationships, and nuances in word meanings. L.7.6 Acquire and use accurately grade-appropriate general academic and domain-specific words and phrases; gather vocabulary knowledge when considering a word or phrase important to comprehension or expression.

Information for the Instructor

What is an Analogy?

An analogy is a comparison between two things. It points out the similarities or likenesses between things that might be different in other circumstances. Analogies form much of the basis of humor, poetry, and metaphors. They draw parallels between the common characteristics of two things and cause us to think analytically about forms, usages, structures and relationships. Analogies are usually written like:

fish : swim :: bird : fly
read as "fish is to swim as bird is to fly"

Analogies for the 21st Century introduces students in intermediate grades to a variety of different types of analogies. Students will learn to solve the analogies by analyzing the relationship between the first two elements of the puzzle and seeking two words that will have the same relationship. Solving analogies builds skills in analytical thinking, flexible thinking and vocabulary.

Structure of the Book

The exercises in this book introduce the different types of analogies by giving examples and then a page of analogies that follow this form. After students have been introduced to several different types of analogies, they are given review lessons that incorporate all of the different types to which they have been introduced.

Point out to students that analogies can be written at least two ways and still maintain the same relationship. The examples that introduce an analogy type will show the analogy written two different ways. For instance, the following analogy relates worker and tools of the trade. It can be written as:

carpenter : hammer :: crossing guard : flag
(worker : tool)
or
hammer : carpenter :: flag : crossing guard
(tool : worker)

When students work through the review lessons, they must be flexible in finding the relationship, but once this relationship is found, they should apply the same format to the second half of the analogy.

Before You Start
Introducing Self-Talk to Solve Analogies

Anyone can improve his or her ability to solve these very simple puzzles. The key that unlocks the door to this improvement is self-talk. When encountering any analogy, instruct students to talk silently to themselves about how they see the relationship between the two items and what the analogy is asking them to compare. Go through several examples orally with students where you model the kind of self-talk they might use to solve the sample analogies. Start by presenting this example:

emu : Australia :: penguin : ___
    a. continent
    b. cold
    c. Antarctica
    d. feathers

After students read the analogy, what happens next is the key. They should think (or say to themselves), "What is the relationship between an emu and Australia?" They know an emu is a large flightless bird and the obvious relationship is "it is indigenous to Australia." They say to themselves, "a penguin is a large flightless bird." Before they ever look at any of the choices that finish the analogy, they remember to say, "An emu comes from Australia and a penguin comes from where?"

DOI: 10.4324/9781003233022-1

Then they are on the right track before they get confused by looking at the options. Their thought process or self-talk is saying, "Where does a penguin live?" Then and only then, they survey the choices given. As you look at the choices as a group, talk about each option and why each one is not the correct option, finally selecting "Antarctica" as the correct answer.

Go through several other examples with your class, both before beginning this unit of study and at various times as you work through the lessons in this book. Some pages have reminders to students to use self-talk to analyze each analogy. This technique will greatly improve their problem-solving ability and their ability to analyze their own thinking processes.

When and How to Guess?

After you have finished the first couple of lessons, you may wish to take time to refine students' strategies for solving analogies. Guessing is not a carte blanche to take avoidable risks. If students must guess, however, the following guidelines are helpful. Go through the strategies one by one with students, giving examples and discussing options.

1. Know the meaning of the words.

One stumbling block that bothers many of the students I talk with is vocabulary. Sometimes they don't honestly know the meaning of the words in an analogy. Well, of course, they can rectify this situation by looking up the meaning(s) of any of the words in question. More students tend not to understand the meaning of words than to not understand the relationship factor of analogies. For example, if they encountered this analogy:

dressmaker : tape measure :: surveyor: __
   a. transit
   b. survivor
   c. Washington
   d. hunter

They probably know the meaning of most of these words. However, they may not be familiar with a "transit" which is a tool often used by a surveyor. Under ordinary circumstances they would just look this word up in the dictionary. If they are doing a lesson or taking a test this may not be an option.

It is also important to remember that many words have more than one meaning. Many can be used as more than one part of speech. For example if you encounter a word such as "glue" or "bloom" or "pledge" in an analogy, it is of utmost importance to determine if it is used as a noun or a verb.

Guessing is never an option, if other means of determination are available. However, guessing can be a useful tool in eliminating choices. Often half of the options can be eliminated quickly.

2. Look for the best choice.

Often students will argue with the makers of analogies. I personally will accept deviations from my answer if a student explains a valid rationale behind his choice. This doesn't mean I will accept any answer or that if you argue with your teacher he or she will be convinced. But on occasion, a creative thinker shows me a whole new way of looking at an analogy. I, therefore, spend more time in sifting out any ambiguity that a student might find. For example:

oats : rice :: polar bear : ___
   a. North Pole
   b. South Pole
   c. seals
   d. Kodiak bear

If you were doing your self-talk, you would probably say something like oats and rice are both specific types of grains. Now you come to polar bear. You might say a polar bear is a specific animal or a specific mammal or a specific mammal that lives in a cold climate. If you say this, then it would seem that both option c. and d. are correct complements for this analogy. One option, however, is superior to the other. Probably the analogy maker would like you to pick Kodiak bear, since these are two specific types of bears.

3. Pick an option that is parallel.

This simply means if an analogy is presented as noun : verb :: noun : ___, then you are seeking a verb. If, on the other hand, you encounter adjective : noun :: adjective : ___, then you are seeking a noun. An example would be:

potpourri : fragrant :: music : ___
   a. notes
   b. tuba
   c. melodious
   d. lesson

Students may not know what the word "potpourri" means but they realize fragrant is an adjective. Therefore to keep everything parallel they would seek an adjective, a word that somehow modifies the word music. Option c. is the only adjective.

4. Guess sensibly after you have narrowed down your odds.

Encourage students to apply everything they have learned including self-talk, parallelism, and meaning. Doing this often eliminates at least half of the options. They should then ask, "Which is the best choice?"

Self-Talk to Improve Your Analogy Abilities

Before you start studying analogies, you need to practice a strategy that will help you solve these puzzles. This strategy is **self-talk**. When solving any analogy, talk silently to yourself about what the relationship between the two items is and what the analogy is asking you to compare. Here is an example.

emu : Australia :: penguin : ____
   1. continent
   2. cold
   3. Antarctica
   4. feathers

⇨ First say to yourself, **"What is the relationship between an emu and Australia?"**
   You know an emu is a large flightless bird and the obvious relationship is **"it is indigenous to Australia."**

⇨ Before you look at any of the choices that finish the analogy, say to yourself, **"An emu comes from Australia and a penguin comes from where?"** You are already on the right track before you confuse yourself by looking at the options. You are saying, **"Where does a penguin live?"**

⇨ Then, and only then, do you survey the choices.

   1. <u>continent</u> - This is a place but not a specific place.

   2. <u>cold</u> - This only describes a place.

   3. <u>Antarctica</u> - This is a place, and penguins live here. This is probably the correct choice, but check out the next option before you decide.

   4. <u>feathers</u> - These are a part of a penguin but not a place.

⇨ The correct answer is **Antarctica**.

Try this second example. Follow the self-talk as you work through the analogy.

fin : fish :: bill : ___
   1. receipt
   2. bird
   3. wing
   4. eel

1. A fin is a part of a fish. A bill is a part of what?

2. I am looking for an animal with a bill.

3. "Receipt" and "foot" are not animals.

4. Bird and eel are animals, but eels don't have bills.

5. The correct answer is bird. Boy, am I smart!.

© Taylor & Francis • *Analogies for the 21st Century*
DOI: 10.4324/9781003233022-2

Introducing   synonyms

pity : sympathy :: catch : capture

antonyms

empty : full :: cheap : expensive

■■

1. dawn : sunrise :: dusk : __
a. dust
b. musk
c. sunset
d. Dutch

2. AM : PM :: day : __
a. time
b. clock
c. night
d. hourglass

3. jawbone : mandible :: backbone : __
a. spine
b. fish
c. animal
d. calcium

4. good : evil :: noon : __
a. midday
b. high
c. midnight
d. lunch

5. a : z :: alpha : __
a. omega
b. Roman
c. Greek
d. alphabet

6. winter : summer :: west : __
a. western
b. California
c. east
d. south

7. landing : take off :: entrance : __
a. door
b. emergency
c. exit
d. enter

8. bravery : cowardice :: jumbo : __
a. elephant
b. minuscule
c. colossal
d. jumble

9. style : fashion :: dog : __
a. house
b. food
c. canine
d. collar

10. polygraph : lie detector :: perjury : __
a. false testimony
b. juror
c. jury
d. witness

11. flag : banner :: lad : __
a. ladder
b. boy
c. stripes
d. stars

12. cash : money :: sofa : __
a. soft
b. cushions
c. couch
d. chair

13. Polaris : North Star :: bison : __
a. mammal
b. buffalo
c. constellation
d. plains

14. sunrise : sunset :: morning : __
a. mourning
b. evening
c. coffee
d. breakfast

© Taylor & Francis • *Analogies for the 21st Century*

DOI: 10.4324/9781003233022-3

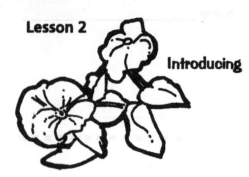

Lesson 2

Name _____

Introducing

whole : part
flower : petal :: foot : toe
petal : flower :: toe : foot

general : specific
bird : meadowlark :: dessert : gingerbread
meadowlark : bird :: gingerbread : dessert

■■

1. Cheddar : cheese :: oats : __
a. wheat
b. rice
c. grain
d. soybeans

2. Monday : day :: January : __
a. winter
b. cold
c. month
d. February

3. fir : evergreen :: oak : __
a. forest
b. deciduous
c. acorn
d. lumber

4. Appaloosa : horse :: Jersey : __
a. New Jersey
b. cow
c. cream
d. jargon

5. blackberry cobbler : dessert :: clam chowder : __
a. shells
b. soup
c. tureen
d. ladle

6. toe : toenail :: finger : __
a. file
b. polish
c. manicure
d. fingernail

7. Potomac : river :: Pacific : __
a. calm
b. Atlantic
c. ocean
d. gigantic

8. blue jay : bird :: trout : __
a. stream
b. broiled
c. fish
d. gills

9. propeller : airplane :: wick : __
a. wicker
b. candle
c. burn
d. helicopter

10. butterfly : monarch :: snake : __
a. copperhead
b. coil
c. strike
d. lizard

11. board game : Monopoly :: cat : __
a. Siamese
b. kitten
c. claws
d. litter

12. Nashville : Tennessee :: city : __
a. town
b. county
c. capital
d. state

13. lens : microscope :: star : __
a. shine
b. shiny
c. constellation
d. comet

14. knife : blade :: squirrel : __
a. nut
b. tail
c. chipmunk
c. acorn

8

© Taylor & Francis • *Analogies for the 21st Century*
DOI: 10.4324/9781003233022-4

Introducing   **thing : characteristic**
fire : hot :: candy : sweet
hot : fire :: sweet : candy
group : member
jury : juror :: flock : goose
juror : jury :: goose : flock

■■

1. yellow : canary :: red : __
a. green
b. cardinal
c. reddish
d. ready

2. quesadilla : Mexican ::
   pizza : __
a. crust
b. sauce
c. Italian
d. pepperoni

3. flock : geese :: herd : __
a. heard
b. shepherd
c. goats
d. pasture

4. senator : senate :: tree : __
a. stump
b. leaf
c. deciduous
d. forest

5. potpourri : fragrant ::
   music : __
a. notes
b. tuba
c. melodious
d. lesson

6. bush : hedge :: student : __
a. studious
b. lazy
c. tardy
d. class

7. pillow : soft :: glue : __
a. sticky
b. paper
c. tape
d. paste

8. grapevine : vineyard ::
   apple tree : __
a. apple
b. cider
c. pie
d. orchard

9. lion : pride :: bee : __
a. swarm
b. sting
c. honey
d. wasp

10. flame : brilliant ::
    dungeon : __
a. dank
b. castle
c. prison
d. captives

11. card : deck :: wolf : __
a. carnivorous
b. cub
c. pack
d. coyote

12. plum : juicy :: rain forest : __
a. South America
b. Amazon
c. Brazil
d. humid

13. lace : translucent ::
    enamel : __
a. glossy
b. layers
c. brush
d. nail

14. squadron : planes :: fleet : __
a. fast
b. flight
c. ships
d. float

Remember to use self-talk to analyze the analogies even if they seem so easy you don't think you need it.

© Taylor & Francis • *Analogies for the 21st Century*

DOI: 10.4324/9781003233022-5

Introducing   **thing or place : what it might contain or house**
aquarium : fish :: hive : bee
fish : aquarium :: bee : hive
singular : plural
mouse : mice :: house : houses
mice : mice :: houses : house

1. pillow case : pillow ::
   envelope : __
   a. postage
   b. seal
   c. stamp
   d. letter

2. ax : axes :: egg : __
   a. omelet
   b. eggs
   c. yolk
   d. scrambled

3. arrows : quiver :: peas : __
   a. soy
   b. green
   c. pod
   d. beans

4. meat : freezer :: jewelry : __
   a. jewelry box
   b. diamond
   c. valuable
   d. decorative

5. loaf : loaves : leaf : __
   a. autumn
   b. rake
   c. chlorophyll
   d. leaves

6. pennies : piggy bank ::
   sand : __
   a. sandy
   b. grit
   c. sand box
   d. dollars

7. pillow : feathers :: vase : __
   a. glass
   b. flowers
   c. marble
   d. save

8. cousin : cousins :: rose : __
   a. red
   b. bouquet
   c. fragrant
   d. roses

9. suitcase : clothes :: diary : __
   a. daily
   b. weekly
   c. personal thoughts
   d. journal

10. wallet : money :: briefcase : __
   a. change
   b. toys
   c. shoes
   d. documents

11. pennies : penny ::
   cookies : __
   a. bake
   b. dozen
   c. cookie
   d. oatmeal

12. woman : women ::
   pearl : __
   a. pearls
   b. oyster
   c. necklace
   d. string

13. toothpaste : tube ::
   books : __
   a. knowledge
   b. pages
   c. backpack
   d. paper

14. balloon : helium :: car tire __
   a. flat
   b. rubber
   c. steel belted
   d. air

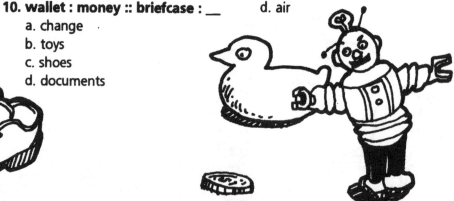

Remember to use self-talk to analyze the analogies even if they seem so easy you don't think you need it.

Lesson 5

Name _____

In this lesson no new analogies will be introduced. Instead you will review types of analogies introduced so far. It is very important to use your self-talk techniques as you compare the relationship between the first two words before trying to select the word that completes the second pair of words.

1. **crime : stock fraud ::**
   university : __
   a. study
   b. professor
   c. Notre Dame
   d. schedule

2. **armada : ship :: swarm : __**
   a. mob
   b. bee
   c. warm
   d. baseball

3. **plant : greenhouse :: ice : __**
   a. cold
   b. frozen
   c. freezer
   d. melt

4. **child : children :: his : __**
   a. hers
   b. theirs
   c. boy
   d. possessive

5. **cookie : gingersnap ::**
   beverage : __
   a. fizz
   b. root beer
   c. pour
   d. pint

6. **Hemingway : writer ::**
   topaz : __
   a. city
   b. gem
   c. cat
   d. South America

7. **lemon : sour :: perfume : __**
   a. fragrant
   b. bottle
   c. cologne
   d. spray

8. **ignite : extinguish ::**
   construct : __
   a. brick
   b. concrete
   c. build
   d. demolish

9. **nocturnal : owl ::**
   carnivorous : __
   a. omnivorous
   b. meat
   c. eagle
   d. pigeon

10. **lion : pride :: ant : __**
    a. cooperative
    b. colony
    c. insect
    d. spider

11. **chief : chiefs :: calf : __**
    a. pasture
    b. young
    c. calves
    d. cows

12. **wealthy : rich :: income : __**
    a. expense
    b. debt
    c. budget
    d. revenue

13. **thesaurus : synonym ::**
    dictionary : __
    a. look up
    b. definition
    c. heavy
    d. lexicographer

14. **whales : pod :: fish : __**
    a. fishes
    b. cold-blooded
    c. ocean
    d. school

Identify each analogy as to its correct type. Choose only one answer.

1. cocoon : butterfly
a. part : whole
b. characteristic : thing
c. group : member
d. container : contents

2. ruby : gem
a. contents :container
b. part : whole
c. group : member
d. member : group

3. city : Milwaukee
a. part : whole
b. general : specific
c. specific : general
d. thing : characteristic

4. cat : feline
a. antonyms
b. general : specific
c. container : contents
d. synonyms

5. meal : breakfast
a. antonyms
b. general : specific
c. container : contents
d. synonyms

6. suppress : restrain
a. synonyms
b. antonyms
c. whole : part
d. thing : characteristic

7. petal : flower
a. container : contents
b. contents : container
c. whole : part
d. part : whole

8. hamper : laundry
a. characteristic : thing
b. group : member
c. member : group
d. container : contents

9. ice : cold
a. thing : characteristic
b. group : member
c. whole : part
d. thing : contents

10. loathe : love
a. antonyms
b. whole : part
c. specific : general
d. thing : characteristic

11. beaker : acid
a. synonyms
b. antonyms
c. container : contents
d. contents : container

12. knife : sheath
a. container : contents
b. contents : container
c. whole : part
d. part : whole

13. spectator : observer
a. synonyms
b. whole : part
c. part : whole
d. specific : general

14. solemn : carefree
a. whole : part
b. antonym
c. thing : contents
d. synonym

15. gold : precious
a. synonym
b. part : whole
c. general : specific
d. thing : characteristic

Remember to use self-talk to analyze the analogies even if they seem so easy you don't think you need it.

© Taylor & Francis • *Analogies for the 21st Century*
DOI: 10.4324/9781003233022-8

Lesson 7

Introducing   **worker : tool**
mechanic : wrench :: teacher: chalk
wrench : mechanic :: chalk : teacher
two members of the same class
cashew : walnut :: salamander : frog

1. spearmint : peppermint : :
   kiwi : __
   a. strawberry
   b. green
   c. juicy
   d. ripe

2. clam : oyster :: chowder : __
   a. bisque
   b. ocean
   c. shells
   d. New England

3. vanilla : chocolate :: collie: __
   a. coat
   b. Lassie
   c. cocker spaniel
   d. fleas

4. racket : tennis player ::
   oven : __
   a. baker
   b. cookie
   c. hot
   d. microwave

5. fireman : hose :: surgeon : __
   a. doctor
   b. cut
   c. scalpel
   d. incision

6. blacksmith : anvil ::
   accountant : __
   a. calculator
   b. calculate
   c. account
   d. count

7. blow dryer : hair stylist ::
   whistle : __
   a. shrill
   b. shiny
   c. lifeguard
   d. blow

8. poison ivy : poison oak ::
   black-eyed Susan : __
   a. sunflower
   b. sun
   c. grow
   d. oil

9. oats : rice :: Polar bear : __
   a. North Pole
   b. South Pole
   c. Santa Claus
   d. Kodiak bear

10. dime : penny :: Tuesday : __
   a. special
   b. Monday
   c. October
   d. week

11. dressmaker : tape measure ::
   surveyor : __
   a. transit
   b. survivor
   c. Washington
   d. hunter

12. artist : palette :: sculptor : __
   a. sculpture
   b. chisel
   c. gallery
   d. creative

13. wren : finch :: salmon : __
   a. stream
   b. swim
   c. ocean
   d. trout

14. Peru : Argentina :: topaz: __
   a. ring
   b. locket
   c. pizza
   d. amethyst

Name _____

Introducing

thing : what it does
duck : migrate :: bear : hibernate
migrate : duck :: hibernate : bear

thing : product
sunflower : oil :: grape : wine
oil : sunflower :: wine : grape

1. tornado : swirl :: volcano : __
a. mountain
b. lava
c. erupt
d. volcanic ash

2. rotate : carousel :: grind : __
a. mill stone
b. ground
c. flour
d. corn meal

3. maple : syrup :: cacao :
a. jungle
b. bean
c. chocolate
d. tapioca

4. earth : rotate :: moon : __
a. satellite
b. revolve
c. crater
d. celestial

5. burrow : ground squirrel :: tunnel : __
a. mole
b. bridge
c. narrow
d. funnel

6. horse : whinny :: goat : __
a. bleat
b. beard
c. nanny
d. billy

7. plum : prune :: tomato : __
a. vegetable
b. seeds
c. red
d. ketchup

8. ice : water :: water : __
a. liquid
b. H$_2$O
c. glass
d. steam

9. tree : lumber :: flax : __
a. linen
b. lax
c. relax
d. soft

10. changes color : chameleon :: plays dead : __
a. quiet
b. still
c. predator
d. opossum

11. silkworm : silk :: spider : __
a. black widow
b. cob web
c. glass
d. tarantula

12. cut : scissor :: measure : __
a. foot
b. distance
c. ruler
d. stethoscope

13. heart : pumps :: teeth : __
a. tooth
b. chew
c. gums
d. mouth

14. pickle : cucumber :: lumber : __
a. plywood
b. house
c. tree
d. planks

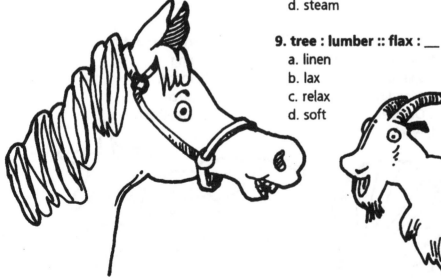

© Taylor & Francis • *Analogies for the 21st Century*
DOI: 10.4324/9781003233022-10

Introducing   **parts of a hierarchy**
freshman : sophomore :: cub scout : boy scout
instrument : thing being measured
hour glass : time :: ruler : distance
time : hour glass :: distance : ruler

■ ■

1. penny : dime :: dime : __
a. silver
b. dollar
c. coin
d. money

2. early childhood : adolescence :: sophomore : __
a. high school
b. class
c. tenth
d. junior

3. air pressure : tire gauge :: pH : __
a. litmus paper
b. light
c. color
d. length

4. clock : time :: balance scale : __
a. weight
b. volume
c. pressure
d. length

5. test : knowledge :: barometer : __
a. oil pressure
b. blood pressure
c. tire pressure
d. atmospheric pressure

6. week : month :: month : __
a. September
b. day
c. year
d. calendar

7. gram : milligram :: meter : __
a. length
b. kilometer
c. tiny
d. millimeter

8. gasoline : fuel gauge :: visual acuity : __
a. glasses
b. contact lens
c. eye chart
d. 20/20

9. understudy : starring lead :: applause : __
a. audience
b. theater
c. standing ovation
d. playwright

10 breeze : gale :: gale : __
a. breeze
b. tempest
c. blow
d. town

11. page : squire :: squire : __
a. crusade
b. horse
c. knight
d. castle

12. village : town :: town : __
a. house
b. office
c. skyline
d. city

13. kilometer : length :: kilogram : __
a. capacity
b. gram
c. weight
d. candy

14. cup : pint :: inch : __
a. measure
b. sewing
c. foot
d. pinch

Remember to use self-talk to analyze the analogies even if they seem so easy you don't think you need it.

© Taylor & Francis • *Analogies for the 21st Century*

15

DOI: 10.4324/9781003233022-11

In this lesson no new analogies will be introduced. Instead you will review types of analogies that you have already learned how to solve. It is very important to think of the relationship between the first two words before you try to select the word that completes the second pair of words.

1. diamonds : sapphires ::
Panama : __
a. Pacific Ocean
b. canal
c. country
d. Suez

2. abase : exalt :: comply : __
a. apply
b. rebel
c. compliance
d. easy

3. flirtation : coquetry ::
promise : __
a. pledge
b. friends
c. loyal
d. feline

4. soup : bowl :: coffee : __
a. cup
b. hot
c. saucer
d. champagne

5. dentist : drill :: farmer : __
a. industrious
b. cultivate
c. plow
d. crop

6. Mount McKinley : mountain
:: mallard : __
a. marmalade
b. duck
c. pond
d. nest

7. bone : clavicle :: joint : __
a. squeaky
b. movement
c. elbow
d. joist

8. Superman : super hero ::
tuna : __
a. Atlantic
b. Pacific
c. fish
d. net

9. tax : taxes :: datum : __
a. calculation
b. statistics
c. facts
d. data

10. Aztec : Inca :: France : __
a. culture
b. French
c. Europe
d. Spain

11. obey : disobey :: docile : __
a. greedy
b. rebellious
c. hungry
d. water

12. meter : distance :: calorie : __
a. weight
b. height
c. heat
d. kilogram

13. seamstress : sewing machine
:: photographer : __
a. artist
b. album
c. pictures
d. camera

14. vase : flowers :: teapot : __
a. spout
b. China
c. lid
d. tea

Remember to use self-talk to analyze the analogies even if they seem so easy you don't think you need it.

© Taylor & Francis • *Analogies for the 21st Century*
DOI: 10.4324/9781003233022-12

In this lesson no new analogies will be introduced. Instead you will review types of analogies that you have already learned how to solve. It is very important to think of the relationship between the first two words before you try to select the word that completes the second pair of words.

1. plunge : dive ::
   carte blanche : __
   a. cartographer
   b. full authority
   c. pony cart
   d. topography

2. affectionate : sentimental ::
   veto : __
   a. President
   b. Congress
   c. Vice President
   d. negate

3. sink : rise :: exorbitant : __
   a. cheap
   b. shopping
   c. credit
   d. dues

4. vertical : horizontal ::
   latitude : __
   a. friendly
   b. hostility
   c. longitude
   d. numerous

5. bee : fly :: frog : __
   a. jump
   b. pond
   c. amphibian
   d. toad

6. grape : raisin :: carbon : __
   a. element
   b. dioxide
   c. graphite
   d. black

7. shoe : sandal ::
   newspaper : __
   a. tabloid
   b. journalist
   c. newsprint
   d. delivery

8. grape : vine :: acorn : __
   a. walnut
   b. hickory nut
   c. oak tree
   d. squirrel

9. nation : South Korea ::
   dinosaur : __
   a. extinct
   b. prehistoric
   c. stegosaurus
   d. reptile

10. tear : tears :: knife : __
    a. sharp
    b. steel
    c. blade
    d. knives

11. painting : Mona Lisa ::
    building : __
    a. marble
    b. Taj Mahal
    c. temple
    d. architecture

12. bone : mandible :: musical
    note : __
    a. quarter note
    b. staff
    c. music
    d. trombone

13. dishwasher : dishes ::
    autoclave : __
    a. cars
    b. surgical instruments
    c. cave
    d. sterile

14. bee : wasp :: dolphin : __
    a. smart
    b. swim
    c. ocean
    d. porpoise

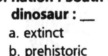

Identify each analogy as to its correct type. Choose only one answer.

■ ■

1. hutch : rabbits
   a. tool : worker
   b. container : contents
   c. contents : container
   d. thing : characteristic

2. corridor : hall
   a. thing: product
   b. product: thing
   c. synonyms
   d. singular: plural

3. escape : imprison
   a. member : group
   b. antonyms
   c. container : contents
   d. contents : container

4. unite : divide
   a. member : group
   b. antonyms
   c. container : contents
   d. contents : container

5. centrifuge : scientist
   a. tool : worker
   b. container : contents
   c. synonym
   d. person : area of interest

6. Ernest Hemingway :
   William Faulkner
   a. two members of the same
      class
   b. thing : something it does
   c. characteristic : thing
   d. antonyms

7. long jump : track meet
   a. group : member
   b. part : whole
   c. singular : plural
   d. thing : action

8. ladder : climb
   a. thing : action
   b. characteristic : thing
   c. member : group
   d. container : contents

9. protective : sunglasses
   a. two members of the same
      class
   b. thing: what it does
   c. characteristic : thing
   d. antonyms

10. mysteries : fiction
   a. group : member
   b. member : group
   c. container : contents
   d. contents : container

11. egg : eggs
   a. synonyms
   b. singular : plural
   c. plural : singular
   d. parts of a hierarchy

12. nectarine : fruit
   a. specific : general
   b. thing : characteristic
   c. characteristic : thing
   d. group : member

13. Africa : Australia
   a. whole : part
   b. two members of the same
      class
   c. hierarchy
   d. thing : contents

14. creek : river
   a. thing : characteristic
   b. hierarchy
   c. thing : action
   d. thing : what it does

15. summit : mountain
   a. part : whole
   b. two members of the same
      class
   c. synonym
   d. general : specific

© Taylor & Francis • *Analogies for the 21st Century*
DOI: 10.4324/9781003233022-14

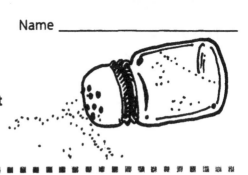

Introducing   **things that go together**
salt : pepper :: knife : fork
foundations : things they support
root : tree :: base : pyramid
tree : root :: pyramid : base

---

1. dawn : rooster :: dusk : __
   a. evening
   b. whippoorwill
   c. bicycle
   d. bog

2. congress : legislative ::
   president : __
   a. vice president
   b. White House
   c. executive
   d. olive branch

3. cactus : desert :: orchid : __
   a. purple
   b. vanilla
   c. rain forest
   d. hybrid

4. statue : pedestal ::
   birdhouse : __
   a. nest
   b. pole
   c. blue bird
   d. sparrow

5. key : lock :: smoke : __
   a. haze
   b. fire
   c. smoky
   d. hot

6. Romeo : Juliet :: comb : __
   a. play
   b. Shakespeare
   c. tooth
   d. brush

7. needle : thread :: hook : __
   a. curved
   b. snag
   c. fish
   d. fishing line

8. tree : trunk :: flower : __
   a. pink
   b. rose
   c. stem
   d. bee

9. golf ball : tee :: ice cream : __
   a. vanilla
   b. chocolate
   c. cone
   d. sherbet

10. pyramid : base :: stool : __
   a. wood
   b. hard
   c. legs
   d. ladder

11. Hansel : Gretel :: tortoise : __
   a. turtle
   b. hare
   c. slow
   d. reptile

12. telescope : tripod :: car : __
   a. sleek
   b. SUV
   c. mileage
   d. tires

13. Mickey : Minnie ::
   Anthony : __
   a. man
   b. boy
   c. girl
   d. Cleopatra

14. paper : pencil :: block : __
   a. cube
   b. wood
   c. tackle
   d. wheel

Remember to use self-talk to analyze the analogies even if they seem so easy you don't think you need it.

Name _____

Introducing   **thing : function**
stove : cook :: spoon : stir
cook : stove :: stir : spoon
male : female
ram : ewe :: actor : actress
ewe : ram :: actress : actor

■■■

1. niece : nephew :: aunt : __
a. ant
b. uncle
c. cousin
d. Polly

2. sift : sieve :: whip : __
a. boots
b. potatoes
c. egg whites
d. whisk

3. bell : ring :: wheel : __
a. roll
b. black
c. car
d. wagon

4. goose : gander :: hen : __
a. egg
b. nest
c. coop
d. rooster

5. ram : ewe :: bull : __
a. cow
b. ranch
c. hoof
d. horn

6. bake : oven :: grill : __
a. sauce
b. ribs
c. baste
d. barbeque

7. knife : slice :: solvent : __
a. dissolve
b. solve
c. gasoline
d. harden

8. traffic light : regulate :: escalator : __
a. move
b. elevator
c. staircase
d. up

9. mare : stallion :: doe : __
a. buck
b. wool
c. horse
d. lamb

10. actor : actress :: señor : __
a. man
b. Spanish
c. señora
d. senior

11. kiln : to heat :: needles : __
a. to knit
b. steel
c. sharp
d. yarn

12. dictionary : define :: compass : __
a. magnetic
b. direct
c. north
d. ship

13. Mrs. : Mr. :: her : __
a. him
b. mine
c. their
d. we

14. mouse : computer :: remote control : __
a. lost
b. buttons
c. television
d. defective

© Taylor & Francis • *Analogies for the 21st Century*
DOI: 10.4324/9781003233022-16

Name _____

Introducing   thing : outer covering
   pie : crust :: banana : peel
   crust : pie :: peel : banana
thing : top
   skyscraper : penthouse :: sundae : cherry
   penthouse : skyscraper :: cherry : sundae

1. sheet : mattress ::
   frosting : __
   a. icing
   b. cake
   c. butter
   d. spread

2. house : roof :: fireplace : __
   a. screen
   b. fire
   c. mantle
   d. hearth

3. cytoplasm : cell wall ::
   melon : __
   a. patch
   b. vine
   c. cucumber
   d. rind

4. fairway : grass :: bed : __
   a. soft
   b. blanket
   c. king size
   d. head board

5. nut : shell :: corn : __
   a. vegetable
   b. cob
   c. husk
   d. kernel

6. spaghetti : sauce ::
   sundae : __
   a. cold
   b. sweet
   c. dessert
   d. whipped cream

7. thimble : finger :: helmet : __
   a. head
   b. spikes
   c. touchdown
   d. pass

8. present : wrapping paper ::
   letter : __
   a. message
   b. envelope
   c. postage
   d. note

9. dog : fur :: fish : __
   a. octopus
   b. hook
   c. scales
   d. tuna

10. human : skin :: duck : __
   a. pond
   b. feathers
   c. quack
   d. drake

11. egg : shell :: book : __
   a. library
   b. pages
   c. cover
   d. tablet

12. flag : flag pole :: chimney : __
   a. smoky
   b. sparks
   c. brick
   d. fireplace

13. Christmas tree : star ::
   church : __
   a. cathedral
   b. building
   c. steeple
   d. religion

14. lid : pot :: cork : __
   a. can
   b. bark
   c. float
   d. bottle

Introducing

worker : product
printer : newspaper :: weaver : cloth
newspaper : printer :: cloth : weaver

symbol : thing represented
peace : olive branch :: good luck : rabbit's foot
olive branch : peace :: rabbit's foot : good luck

worker : job location
doctor : hospital :: logger : forest
hospital : doctor :: forest : logger

---

1. Wall Street : broker :: Nashville : __
a. country songwriter
b. Tennessee
c. skyscrapers
d. cactus

2. portrait : painter :: bust : __
a. vase
b. dust
c. sculptor
d. break

3. ☁ : weather :: ♫ : __
a. joined
b. singer
c. two
d. music

4. dollar : $ :: cent : __
a. nickel
b. dime
c. quarter
d. ¢

5. poem : poet :: cabinet : __
a. shelf
b. shelves
c. carpenter
d. kitchen

6. picture : photographer :: novel : __
a. story
b. writer
c. artist
d. librarian

7. # : number :: % : __
a. hopeful
b. careful
c. accountant
d. per cent

8. − : + :: subtract : __
a. math
b. subtraction
c. reductions
d. add

9. baker : bakery :: waiter : __
a. table
b. dishes
c. food
d. restaurant

10. architect : blueprints :: baker : __
a. oven
b. bakery
c. pastries
d. chef

11. forest : ranger :: swimming pool : __
a. weather forecaster
b. summer
c. swimmers
d. lifeguard

12. hospital : doctor :: lab : __
a. laboratory
b. sanitary
c. scientist
d. microscope

13. teller : bank :: pilot : __
a. flying
b. airplane
c. navigate
d. aviator

14. ☠ : poison :: ♥ : __
a. holiday
b. circulation
c. lonely
d. love

Remember to use self-talk to analyze the analogies even if they seem so easy you don't think you need it.

© Taylor & Francis • *Analogies for the 21st Century*
DOI: 10.4324/9781003233022-18

Lesson 17

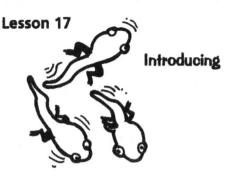

Name _____

Introducing

person : area of interest
veterinarian : animals :: paleontologist : fossils
animals : veterinarian :: fossils : paleontologist

thing : what it might become
peat : coal :: sand : glass
coal : peat :: glass : sand

---

1. **tree resin : amber ::**
   carbon : __
   a. car
   b. bonbon
   c. diamond
   d. burns

2. **bud : blossom :: seed : __**
   a. soil
   b. water
   c. cultivate
   d. seedling

3. **song writer : music ::**
   choreographer : __
   a. chorus
   b. dance
   c. instructor
   d. art

4. **steel : iron :: brass : __**
   a. metal
   b. band
   c. copper
   d. shiny

5. **nurse : medical aid ::**
   broker : __
   a. financial advice
   b. broken
   c. breaking
   d. broke

6. **calf : cow :: fingerling : __**
   a. glove
   b. fish
   c. young
   d. hand

7. **glasses : optician ::**
   computer : __
   a. computations
   b. keyboard
   c. software engineer
   d. Internet

8. **chef : food :: florist : __**
   a. flowers
   b. arrange
   c. fauna
   d. shop

9. **entrepreneur : business ::**
   politician : __
   a. policy
   b. politics
   c. campaign
   d. corrupt

10. **lumber : building :: flour : __**
    a. wheat
    b. bread
    c. yeast
    d. white

11. **tadpole : frog :: pupa : __**
    a. toad
    b. change
    c. puppy
    d. caterpillar

12. **architect : buildings ::**
    astronomer : __
    a. celestial bodies
    b. telescope
    c. star gazer
    d. observatory

13. **Morse : telegraph ::**
    Edison : __
    a. inventor
    b. electric light
    c. American
    d. laboratory

14. **gosling : goose :: cygnet : __**
    a. graceful
    b. fowl
    c. swan
    d. gander

© Taylor & Francis • *Analogies for the 21st Century*

DOI: 10.4324/9781003233022-19

In this lesson no new analogies will be introduced. Instead you will review types of analogies that you have already learned how to solve. It is very important to think of the relationship between the first two words before you try to select the word that completes the second pair of words.

1. money bag : money ::
   briefcase : __
   a. documents
   b. erase
   c. open
   d. spill

2. rehearse : practice ::
   violate : __
   a. purple
   b. transgress
   c. allow
   d. blossom

3. dutiful : obedient :: yield : __
   a. ripe
   b. open
   c. surrender
   d. cork

4. jeweler : loupe ::
   acupuncturist : __
   a. cure
   b. treat
   c. needle
   d. pain

5. Elizabeth I : queen ::
   Shakespeare : __
   a. Avon
   b. Globe Theater
   c. English
   d. playwright

6. explorer : DeSoto ::
   general : __
   a. army
   b. command
   c. MacArthur
   d. ★★★★★

7. rung : ladder :: clapper : __
   a. clap
   b. claptrap
   c. bell
   d. clapboard

8. tooth : teeth :: deer : __
   a. doe
   b. buck
   c. fawn
   d. deer

9. canopy : bed :: awning : __
   a. window
   b. cover
   c. striped
   d. rain

10. cup : pint :: quart : __
   a. milk
   b. paint
   c. water
   d. gallon

11. Rio Grande : Yangtze ::
   English spaniel : __
   a. dog
   b. canine
   c. golden retriever
   d. pet

12. spectacles : eyeglasses ::
   walking stick : __
   a. rims
   b. cane
   c. vision
   d. perambulate

13. ™ : trademark :: © : __
   a. circle C
   b. register
   c. symbol
   d. copyright

14. Cousteau : marine life ::
   Audubon : __
   a. automation
   b. insects
   c. birds
   d. auditory

Remember to use self-talk to analyze the analogies even if they seem so easy you don't think you need it.

© Taylor & Francis • *Analogies for the 21st Century*
DOI: 10.4324/9781003233022-20

In this lesson no new analogies will be introduced. Instead you will review types of analogies that you have already learned how to solve. It is very important to think of the relationship between the first two words before you try to select the word that completes the second pair of words.

■■■

1. **direction : ➡ ::**
   quotation : __
   a. ♪
   b. 66
   c. saying
   d. ☆

2. **tree : sycamore :: duck : __**
   a. decoy
   b. hunting
   c. wetlands
   d. Muscovy

3. **flashlight : battery :: car : __**
   a. automobile
   b. cars
   c. fleet
   d. engine

4. **heroes : hero :: bones : __**
   a. skeleton
   b. bone
   c. broken
   d. doctor

5. **offense : defense :: + : __**
   a. &
   b. plus
   c. —
   d. addition

6. **teacher : classroom ::**
   lawyer : __
   a. judge
   b. attorney
   c. courtroom
   d. litigate

7. **change : amend ::**
   improvement : __
   a. government
   b. president
   c. document
   d. betterment

8. **pantry : food :: vase : __**
   a. jug
   b. glass
   c. bouquet
   d. valise

9. **extrovert : gregarious ::**
   introvert : __
   a. shy
   b. jerk
   c. hermit
   d. concave

10. **cottage : house ::**
    rowboat : __
    a. yacht
    b. oar
    c. float
    d. lake

11. **eaglet : eagle :: filly : __**
    a. mare
    b. frilly
    c. foal
    d. mammal

12. **cotton : cloth :: coffee : __**
    a. brown
    b. roasted
    c. Brazil
    d. beverage

13. **hoodwink : mislead ::**
    revolt : __
    a. revolting
    b. rebel
    c. revolver
    d. voltage

14. **male : female :: king : __**
    a. authority
    b. prince
    c. princess
    d. queen

15. **wheat : barley :: canoe : __**
    a. flotation
    b. glider
    c. kayak
    d. life jacket

Remember to use self-talk to analyze the analogies even if they seem so easy you don't think you need it.

Name _____

In this lesson no new analogies will be introduced. Instead you will review types of analogies that you have already learned how to solve. It is very important to think of the relationship between the first two words before you try to select the word that completes the second pair of words.

1. emperor : Napoleon ::
playwright : __
a. theater
b. screen writer
c. Shakespeare
d. drama

2. hoe : gardener :: scalpel : __
a. sharp
b. operation
c. knife
d. surgeon

3. incisor : tooth :: Oscar : __
a. actor
b. award
c. Emmy
d. accept

4. store : mall :: house : __
a. subdivision
b. shutters
c. mortgage
d. roof

5. hugs : kisses ::
peanut butter : __
a. George Washington Carver
b. spread
c. sticky
d. jelly

6. tile pieces : mosaic ::
movements : __
a. notes
b. solo
c. musician
d. symphony

7. aquarium : water ::
tanker : __
a. tanks
b. ship
c. spill
d. oil

8. hen : rooster :: mare : __
a. stallion
b. colt
c. horse
d. gallop

9. willing : agreeable ::
motionless : __
a. mobile
b. stationary
c. tree
d. movement

10. gallon : milk :: bushel : __
a. measure
b. peck
c. pint
d. apples

11. birthright : heritage ::
heir : __
a. will
b. money
c. inherit
d. beneficiary

12. ✡ : Jewish :: ✝ : __
a. Christian
b. Buddhist
c. cross
d. religion

13. knife : blade :: daisy : __
a. white
b. dainty
c. flower
d. petal

14. faithful : negligent :: fail : __
a. test
b. exam
c. high school
d. succeed

Remember to use self-talk to analyze the analogies even if they seem so easy you don't think you need it.

© Taylor & Francis • *Analogies for the 21st Century*
DOI: 10.4324/9781003233022-22

In this lesson no new analogies will be introduced be introduced. Instead you will review types of analogies that you have already learned how to solve. It is very important to think of the relationship between the first two words before you try to select the word that completes the second pair of words.

1. law : order :: knife : __
  a. cut
  b. sharp
  c. fork
  d. pocket

2. index : book :: zip code : __
  a. zipper
  b. address
  c. postman
  d. mail box

3. synagogue : rabbi :: church : __
  a. steeple
  b. minister
  c. building
  d. religion

4. bon jour : au revoir :: hello : __
  a. good bye
  b. French
  c. family
  d. bon voyage

5. ✂ : cut :: ✏ : __
  a. write
  b. pencil
  c. pen
  d. calligraphy

6. expensive : costly :: tax : __
  a. pay
  b. overdue
  c. tariff
  d. high

7. calculator : bookkeeper :: grade book : __
  a. names
  b. grades
  c. average
  d. teacher

8. consent : refuse :: permit : __
  a. permission
  b. allow
  c. driving
  d. forbid

9. shell : egg :: armor plates : __
  a. strong
  b. bony
  c. armadillo
  d. road

10. lapel : suit :: cuff : __
  a. iron
  b. laundry
  c. button hole
  d. shirt

11. grave : cemetery :: pane : __
  a. window
  b. panes
  c. pain
  d. broken

12. meek : submissive :: endure : __
  a. duration
  b. pain
  c. bear
  d. meek

13. hate : love :: calm : __
  a. clam
  b. sedate
  c. jittery
  d. exorbitant

14. philatelist : stamps :: numismatist : __
  a. numbers
  b. numerals
  c. coins
  d. magic

In this lesson no new analogies will be introduced. Instead you will review types of analogies that you have already learned how to solve. It is very important to think of the relationship between the first two words before you try to select the word that completes the second pair of words.

■■

1. spring : summer :: fall : __
   a. leaves
   b. winter
   c. October
   d. season

2. garage : car :: closet : __
   a. bedroom
   b. cupboard
   c. house
   d. clothes

3. California : state :: Picasso : __
   a. art
   b. artist
   c. cubes
   d. unusual

4. lice : louse :: dice : __
   a. douse
   b. die
   c. toss
   d. game

5. soap : water :: lightening : __
   a. bright
   b. dangerous
   c. streaks
   d. thunder

6. fire engine : firehouse :: aircraft : __
   a. hangar
   b. passenger
   c. pilot
   d. parachute

7. Jupiter : Neptune :: Iroquois : __
   a. nation
   b. tribe
   c. native American
   d. Cherokee

8. magnet : attract :: glue : __
   a. white
   c. adhere
   d. bottle
   e. sticky

9. hear no evil : see no evil :: monkey see : __
   a. see the monkey
   b. catch the monkey
   c. pin the tail on the monkey
   d. monkey do

10. summit : base :: rise : __
   a. sink
   b. zinc
   c. drink
   d. pink

11. stethoscope : doctor :: telescope : __
   a. archeologist
   b. stars
   c. microscope
   d. astronomer

12. Aesop : fables :: Grimm Brothers : __
   a. fairy tales
   b. German
   c. Hansel and Gretel
   d. writers

13. heed : neglect :: nonchalant : __
   a. high-strung
   b. gone
   c. easy
   d. laundry

14. tail : kite :: sole : __
   a. seafood
   b. fish
   c. shoe
   d. soul

Remember to use self-talk to analyze the analogies even if they seem so easy you don't think you need it.

© Taylor & Francis • *Analogies for the 21st Century*
DOI: 10.4324/9781003233022-24

In this lesson no new analogies will be introduced. Instead you will review types of analogies that you have already learned how to solve. It is very important to think of the relationship between the first two words before you try to select the word that completes the second pair of words.

1. basil : pesto :: avocado : __
   a. guacamole
   b. fruit
   c. dip
   d. garnish

2. husk : corn :: wool : __
   a. knit
   b. yarn
   c. sweater
   d. lamb

3. infatuation : crush ::
   commitment : __
   a. company
   b. obligation
   c. comet
   d. star

4. elevation : depth :: top : __
   a. bottom
   b. base
   c. foundation
   d. basement

5. lavender : purple :: aqua : __
   a. brine
   b. marine
   c. blue
   d. sea

6. pink : color :: sour : __
   a. tart
   b. apple
   c. taste
   d. lemon

7. cover : book :: wallet : __
   a. folded
   b. leather
   c. black
   d. currency

8. salutatorian : valedictorian ::
   lieutenant : __
   a. captain
   b. ambition
   c. ambitious
   d. lazy

9. armor : knight :: cast : __
   a. plaster
   b. gauze
   c. broken arm
   d. sling

10. escape : capture :: prey : __
   a. predator
   b. bobcat
   c. puma
   d. cougar

11. basket : picnic :: hangar : __
   a. closet
   b. airplane
   c. hamburger
   d. ketchup

12. nest : egg :: cradle : __
   a. infant
   b. rock
   c. child
   d. lullaby

13. submissive : insubordinate ::
   over : __
   a. board
   b. cover
   c. bridge
   d. under

14. double-dealing : deceitful ::
   correction : __
   a. error
   b. red
   c. rectification
   d. cheat

In this lesson no new analogies will be introduced. Instead you will review types of analogies that you have already learned how to solve. It is very important to think of the relationship between the first two words before you try to select the word that completes the second pair of words.

1. **banana : peel :: pheasant : __**
   a. hunt
   b. bird
   c. China
   d. feathers

2. **milk : glass :: shampoo : __**
   a. hair
   b. soap
   c. herbal
   d. bottle

3. **piano : key :: family tree : __**
   a. elm
   b. spruce
   c. ancestor
   d. dogwood

4. **stroller : baby :: shopping cart : __**
   a. supermarket
   b. cart boy
   c. parking lot
   d. groceries

5. **ram : ewe :: count : __**
   a. countess
   b. number
   c. duke
   d. sheep

6. **fungus : spore :: evergreen : __**
   a. pine
   b. fir
   c. spruce
   d. cone

7. **bricks : brick :: gloves : __**
   a. leather
   b. golf
   c. glove
   d. mittens

8. **delude : dupe :: cheat : __**
   a. honor
   b. obey
   c. swindle
   c. bargain

9. **diamond : baseball :: course : __**
   a. football
   b. golf
   c. coarse
   d. game

10. **don't : $\bigcirc$ :: greater than : __**
    a. >
    b. <
    c. ∨
    d. ≅

11. **oath : vow :: infringe : __**
    a. fringe
    b. surrey
    c. trespass
    d. rights

12. **friend : foe :: do : __**
    a. due
    b. doe
    c. overdue
    d. don't

13. **Beethoven : music :: Disney : __**
    a. land
    b. entertainment
    c. Walt
    d. Mickey Mouse

14. **rock : roll :: bride : __**
    a. groom
    b. rehearsal
    c. cake
    d. argue

© Taylor & Francis • *Analogies for the 21st Century*
DOI: 10.4324/9781003233022-26

Do-It Yourself

Note to Teachers: The following are sets of words that can be used in analogies. Vary the your study of analogies by giving students these pairs of words and asking them to find two other words to complete the analogy. You should then discuss the various answers and verify that students' analogies do, indeed, use the same relationship as the first two words. Since this is a difficult task, we suggest that you only give students a few analogies to work on at a time.

general - specific
Challenger : space shuttle
wind instrument : flute
mythological creature : Pegasus
museum : Smithsonian
amphibian : frog
planet : Uranus
citrus : tangerine

whole - part
wheel : spoke
clock : hand
plant : root
colony : ant
ship : anchor
schooner : sail
tree : branch
octopus : tentacle

instrument - what it measures
wind sock : air direction
theodolite : angles
clock : time
seismograph : earth movements
anemometer : wind
thermometer : temperature

thing - what it can become
wheat : flour
paper : collage
bud : blossom
solar energy : electricity
wood : paper
iron : steel
hide : leather

thing - characteristic
blade : sharp
textbook : informative
cactus : prickly
rattlesnake : poisonous
Mercury : hot
hummingbird : small

worker - tool
singer : microphone
doctor : stethoscope
dentist : drill
butcher : clever
carpenter : hammer
scientist : microscope

thing : contents
kennel : dog
library : books
camera : film
artery : blood
larvae : cocoon
gallery : art

thing : function
seat belt : restrain
forceps : grip
scale : weigh
light : illuminate
combine : harvest
telescope : magnify
magnet : attract
glue : adhere

things that go together
e-mail : Internet
bread : butter
dollars : cents
hamburger : French Fries
horse : buggy
needle : thread
crime : punishment
baseball : diamond

hierarchy
quarter note : half note
boy : man
mother : grandmother
private : sergeant
pond : lake
cottage : house
vice president : president

male - female
grandma : grandpa
count : countess
duke : duchess
cow : bull
prince : princess

two members of the same class
guitar : zither
octopus : squid
solar : hydroelectric
Mississippi : Amazon
spring : summer
Atlantic : Pacific
peach : apple
microscope : telescope

person : area of interest
entomologist : insects
meteorologist : weather
ornithologist : birds
calligrapher : writing
lawyer : litigation

symbol
flag : country
blue ribbon : the best
gold cup : first place

thing - top or covering
church : steeple
birthday cake : candles
banana: peel
crustacean : shell
bandage : wound

thing : location
iceberg : ocean
oasis : desert
Eiffel Tower : Paris
stalactite : cave

ANSWERS

Lesson 1
1. c
2. c
3. a
4. c
5. a
6. c
7. c
8. b
9. c
10. a
11. b
12. c
13. b
14. b

Lesson 2
1. c
2. c
3. b
4. b
5. b
6. d
7. c
8. c
9. b
10. a
11. a
12. d
13. c
14. b

Lesson 3
1. b
2. c
3. c
4. d
5. c
6. d
7. a
8. d
9. a
10. a
11. c
12. d
13. a
14. c

Lesson 4
1. d
2. b
3. c
4. a
5. d
6. c
7. b
8. d
9. c
10. d
11. c
12. a
13. c
14. d

Lesson 5
1. c
2. b
3. c
4. b
5. b
6. b
7. a
8. d
9. c
10. b
11. c
12. d
13. b
14. d

Lesson 6
1. d
2. d
3. b
4. d
5. b
6. a
7. d
8. d
9. a
10. a
11. c
12. b
13. a
14. b
15. d

Lesson 7
1. a
2. a
3. c
4. a
5. c
6. a
7. c
8. a
9. d
10. b
11. a
12. b
13. d
14. d

Lesson 8
1. c
2. a
3. c
4. b
5. a
6. a
7. d
8. d
9. a
10. d
11. b
12. c
13. b
14. c

Lesson 9
1. b
2. d
3. a
4. a
5. d
6. c
7. d
8. c
9. c
10. b
11. c
12. d
13. c
14. c

Lesson 10
1. d
2. b
3. a
4. a
5. c
6. b
7. c
8. c
9. d
10. d
11. b
12. c
13. d
14. d

Lesson 11
1. b
2. d
3. a
4. c
5. a
6. c
7. a
8. c
9. c
10. d
11. b
12. a
13. b
14. d

Lesson 12
1. b
2. c
3. b
4. b
5. a
6. a
7. b
8. a
9. c
10. b
11. b
12. a
13. b
14. b
15. a

Lesson 13
1. b
2. c
3. c
4. b
5. b
6. d
7. d
8. c
9. c
10. c
11. b
12. d
13. d
14. c

Lesson 14
1. b
2. d
3. a
4. d
5. a
6. d
7. a
8. a
9. a
10. c
11. a
12. b
13. a
14. c

Lesson 15
1. b
2. c
3. d
4. b
5. c
6. d
7. a
8. b
9. c
10. b
11. c
12. d
13. c
14. d

Lesson 16

1. a	8. d
2. c	9. d
3. d	10. c
4. d	11. d
5. c	12. c
6. b	13. b
7. d	14. d

Lesson 17

1. c	8. a
2. d	9. b
3. b	10. b
4. c	11. d
5. a	12. a
6. b	13. b
7. c	14. c

Lesson 18

1. a	8. d
2. b	9. a
3. c	10. d
4. c	11. c
5. d	12. b
6. c	13. d
7. c	14. c

Lesson 19

1. b	8. c
2. d	9. a
3. d	10. a
4. b	11. a
5. c	12. d
6. c	13. b
7. d	14. d
	15. c

Lesson 20

1. c	8. a
2. d	9. b
3. b	10. d
4. a	11. d
5. d	12. a
6. d	13. d
7. d	14. d

Lesson 21

1. c	8. d
2. b	9. c
3. b	10. d
4. a	11. a
5. a	12. c
6. c	13. c
7. d	14. c

Lesson 22

1. b	8. c
2. d	9. d
3. b	10. a
4. b	11. d
5. d	12. a
6. a	13. a
7. d	14. c

Lesson 23

1. a	8. a
2. d	9. c
3. b	10. a
4. a	11. b
5. c	12. a
6. c	13. d
7. d	14. c

Lesson 24

1. d	8. c
2. d	9. b
3. c	10. a
4. d	11. c
5. a	12. d
6. d	13. b
7. c	14. a

T - #0717 - 101024 - C0 - 276/218/2 - PB - 9781593630478 - Gloss Lamination